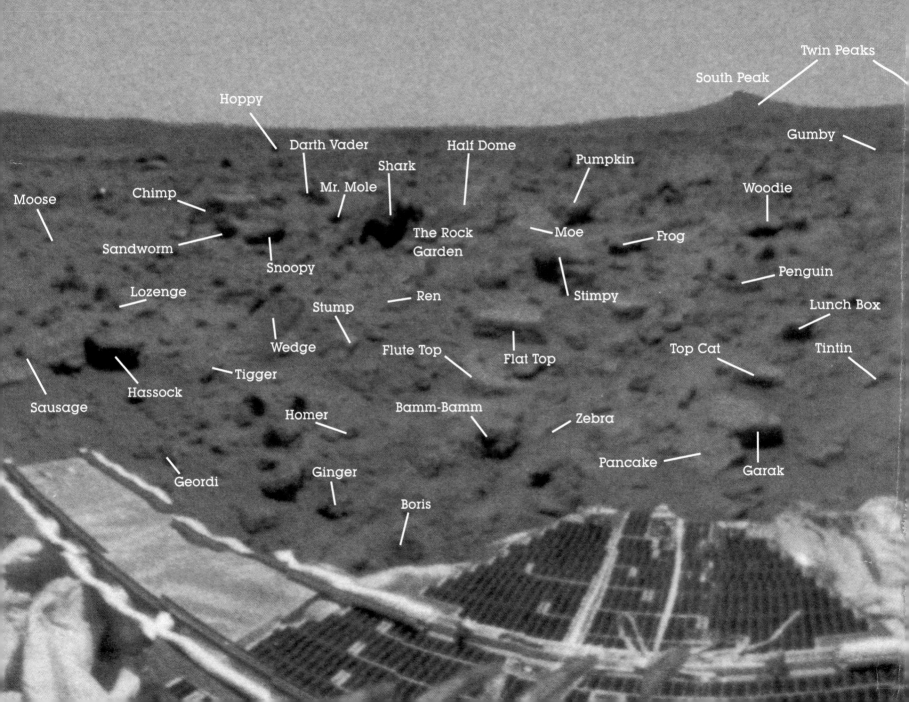

Scientists named the rocks they saw in this Mars panorama, taken on the Pathfinder mission in 1997.

North Peak
Hippo
Potato
Couch
Rolling Stone
Boo Boo
Turtle
Valentine
Flipper
Yogi
Jailhouse
Platypus
Kitten
Pinocchio
Froggy
Jazzy
Warthog
Duck
Cradle
Dragon
Dilbert's Boss
Souffle
Gromit
Ratbert
Pathfinder's Sojourner Rover
Wallace
Barnacle Bill

To Dave and Cari
—F.M.B.

To Sarah Thomson, with thanks
—T.K.

Special thanks to Dave Williams of the
National Space Science Data Center
for his time and expert review

All photographs courtesy of NASA

The *Let's-Read-and-Find-Out Science* book series was originated by Dr. Franklyn M. Branley, Astronomer Emeritus and former Chairman of the American Museum–Hayden Planetarium, and was formerly co-edited by him and Dr. Roma Gans, Professor Emeritus of Childhood Education, Teachers College, Columbia University. Text and illustrations for each of the books in the series are checked for accuracy by an expert in the relevant field. For more information about Let's-Read-and-Find-Out Science books, write to HarperCollins Children's Books, 1350 Avenue of the Americas, New York, NY 10019, or visit our website at www.letsreadandfindout.com.

Library of Congress Cataloging-in-Publication Data
Branley, Franklyn Mansfield, 1915–
    Mission to Mars / by Franklyn M. Branley ; illustrated by True Kelley.
        p.   cm. — (Let's-read-and-find-out science. Stage 2)
    ISBN 0-06-029807-3 — ISBN 0-06-029808-1 (lib. bdg.) —ISBN 0-06-445233-6 (pbk.)
    1. Space flight to Mars—Juvenile literature. (1. Space flight to Mars. 2. Mars (Planet)—Exploration.)
I. Kelley, True, ill. II. Title. III. Series.
TL799.M3 B73 2002   00-054036
629.45'53—dc21   CIP
                                                                                            AC

Typography by Elynn Cohen 1 2 3 4 5 6 7 8 9 10 ❖ First Edition

Mission to MARS

by Franklyn M. Branley • illustrated by True Kelley

Foreword by Neil Armstrong

HarperCollinsPublishers

Foreword

In 1877, an Italian astronomer studied Mars through his telescope. He reported seeing lines on the planet's surface, which he called "canali," meaning grooves or channels. Many people thought this indicated canals constructed by intelligent beings.

Mars and Martians became popular topics of conversation. Fanciful articles, comic strips, short stories, and books were written about Mars and its inhabitants.

During the last forty years, we have sent many space probes to Mars and are now quite certain that there are no canals or inhabitants there. But Mars remains a fascinating place, the most logical new destination for space explorers.

In *Mission to Mars*, you will learn about Mars and how it differs from Earth. You will be able to decide if it is a place you would like to visit. Just possibly, you may have that chance.

—Neil Armstrong

Neil Armstrong commanded the Apollo 11 mission, the spaceflight that landed astronauts on the surface of the moon. He was the first human being to walk on another world.

In the last century, Neil Armstrong became the first person to walk on the Moon. In this century, you may become the first person to walk on Mars.

EARTH

Earth Orbit

International
Space
Station

CREW
TRANSFER
VEHICLE
(CTV)

Your spaceship will be launched
from the International Space Station.
It will take over six months to travel
more than 300 million miles in a
curved path from Earth to Mars.

300 MILLION MILES ⇨

When you reach Mars, you and
the other three people in your
crew will transfer to the Mars
Habitat Lander. It was put into
orbit around Mars months ago.
It will become your station on Mars.

MARS

HABITAT
LANDER

Mars Orbit

CTV

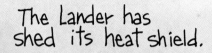
The Lander has shed its heat shield.

You get ready to take the Lander down to the surface. Parachutes open to ease the landing. Engines that slow the ship are fired. They kick up big clouds of dust and powder. After months of space travel, you are on Mars.

The Mars Station where you will live for the next several months is part of the Lander. It has sections for living, sleeping, exercising, and storage, and a laboratory for conducting experiments.

UPPER DECK

LOWER DECK:
Garage,
Workshop,
and Storage Area
Environmental Control
and Life-Support Systems

Air lock

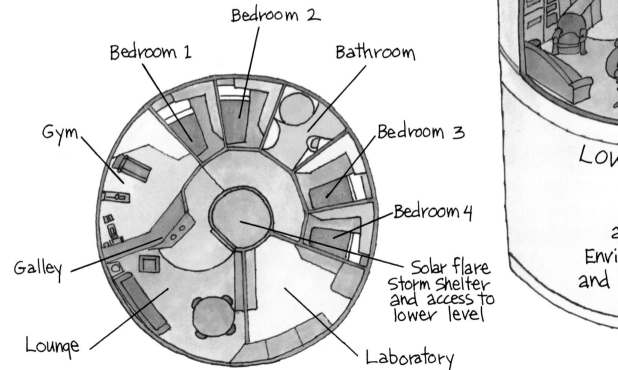

Bedroom 1

Bedroom 2

Bathroom

Gym

Bedroom 3

Bedroom 4

Solar flare storm shelter and access to lower level

Galley

Lounge

Laboratory

Some time ago, cargo ships carrying food, air, and water landed nearby. Your first job is to move some of the supplies from the cargo ships to the Station. Each day the crew uses about 100 pounds of food and water, so tons of it are needed.

Luckily, the gravity on Mars is only one third as strong as the gravity on Earth. The supply packages would be heavy on Earth, but here on Mars you can lift them easily.

This End Up

There is air inside the Station for you to breathe, so you don't need to wear your space suit. Soon you are ready for bed. A Martian day is just a bit over 24 hours long, so the day-night cycle is about the same here as it is on Earth.

At last you can sleep lying down. During the long journey to Mars, you were in microgravity. There was so little gravity that you did not feel it. There was no up or down. On Mars there is enough gravity for you to lie down to sleep without floating away.

After you get up, you check every part of the Station, including the electric system. Electricity is generated by solar panels and an atomic reactor. You make sure that the water recycler is working. Wastewater, water in the air, and urine are purified. They are used over and over again. The air inside the Station is also recycled.

Now you are ready to explore Mars.

Long before telescopes and rockets were invented, people already knew at least one thing about Mars: They knew it is a reddish planet. Red reminded people of blood and war, so the Romans named the planet after Mars, the god of war.

Now we know much more about Mars. We have thousands of pictures of the planet, taken by probes and robots. Mars is as dry and dusty as a desert. But we know that long ago, Mars had water. There are old riverbeds cut into the Martian rock. A huge flat area may have been an ocean.

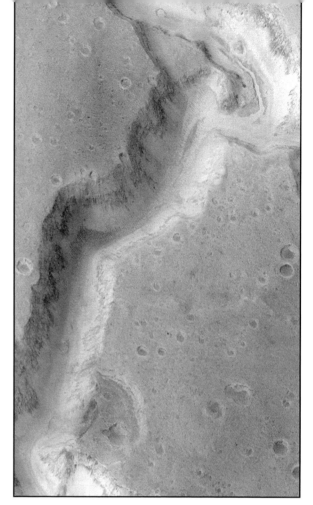

In this picture of a Martian river canyon, the river seems to have changed its path over time.

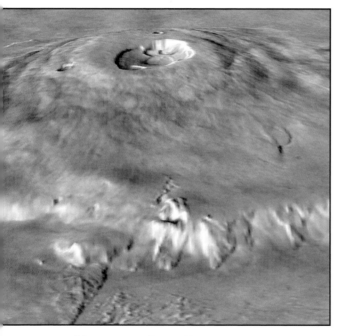

Olympus Mons Volcano is as big as Arizona!

There are many extinct volcanoes. One of them is about 15 miles high. It is the highest peak in the solar system and three times as high as Mount Everest, the highest point on Earth.

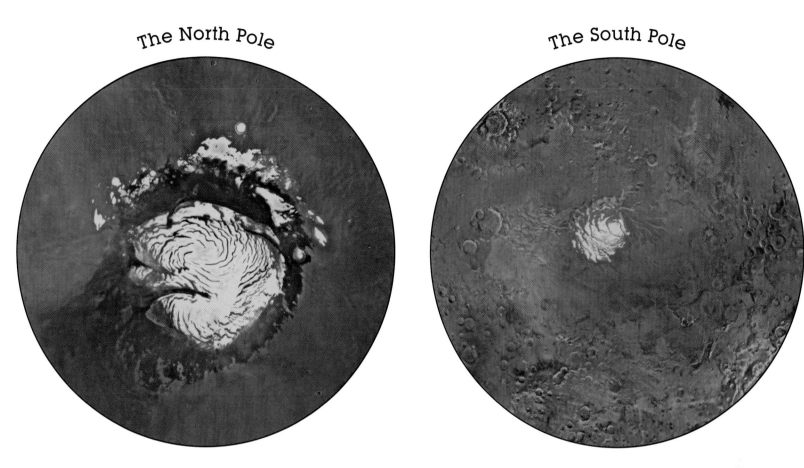

The North Pole

The South Pole

The Martian atmosphere is much thinner than our atmosphere. Most of it is a gas called carbon dioxide. The planet is cold—about 70 degrees Fahrenheit below zero. The North and South Poles are even colder—about 175 degrees Fahrenheit below zero. The poles are covered by ice and by white, frozen carbon dioxide.

It never rains on Mars. There is not enough water in the air.

But it is likely that there is water below the surface of the planet. In some places there is a layer of ice. Below the ice there may be liquid water. It seeps out of the walls of a few craters.

You will dig wells to recover the water. It will be one of your first jobs, because the crew needs a lot of water.

You will also need to grow some of your food. Plants are grown in a wet, spongy material that roots can anchor to. It is moisturized with water containing all the chemicals that plants need.

Your mission to Mars is the first of what may become many more. To help future astronauts, it is important for your crew to discover as much as possible about the planet.

You will probably find no Martian soil. The entire planet seems to be covered by a layer of particles as fine as talcum powder. It comes from meteors that crashed into Mars long ago, and from volcanic explosions that threw lava and dust clouds over the whole planet. The dust blows about and settles on the entire Mars Station. You will have to brush it off the solar collectors and make sure it doesn't get inside the space suits.

There are no plants on Mars—no animals, birds, or bugs. Did they exist there long ago? At some time during the history of Mars, living things may have developed on the planet. You will search for clues that let us know.

You will use a Mars Rover to explore the parts of the planet far away from the Station. You will find craters that were dug out when meteorites crashed into Mars. The planet was bombarded by them billions of years ago. Some of them were huge. One of the craters is at least 200 miles across and so long, it would reach from the east coast of the United States to the Rocky Mountains.

As you move about in the Rover, you will gather samples of rock to be taken back to the laboratory. Very likely, some will be 4 billion or 5 billion years old—as old as the solar system.

When you get to Mars, you will find things no one could have imagined. No one knows what may be discovered, and that is one of the main reasons why we want to go there.

Crew Ascent Vehicle →

After several months of exploration, you get inside the Ascent Vehicle. It was landed on Mars long ago. You fire the engines, and it carries you into Mars orbit.

Then you transfer to the Earth Return Vehicle
that will make the long journey back home.

ERV
(Earth Return
Vehicle)

After the Mars mission, some people dream of traveling out beyond the solar system—all the way to the stars. But stars are very far away. It is 26,000,000,000,000 miles to the nearest one. Before such a journey can be made, we'll need new engines. They will have to travel at tremendous speeds for years and years. It will take at least a century to develop them. Maybe your great-grandchildren will travel to the stars.

You will have to stay closer to Earth. You might become an engineer at the Moon Base or the International Space Station. You might even set out on another mission to Mars.

FIND OUT MORE ABOUT MARS

How much would you weigh on Mars?

If you weigh 60 pounds, that means Earth's gravity is pulling on you with a force of 60 pounds.

Gravity on Mars is only one third of what it is on Earth, so on Mars you would weigh only one third of your Earth weight. If you weigh 60 pounds on Earth, you would weigh 20 pounds on Mars.

If you can lift 25 pounds on Earth, you could lift 75 pounds on Mars. If you can jump 1 foot high on Earth, you could jump 3 feet on Mars.

On Jupiter you would weigh 2.36 times more than you do on Earth. On the Sun, you would weigh 28 times more. On the Moon you would only weigh one sixth of your Earth weight. You could easily jump 6 feet high.

EARTH   MOON   MARS

Life on Mars

Some people think that a long time from now, there will be a colony on Mars. Do you think this could happen? Draw a picture of a Martian colony, or write a story about what a day in a Martian colony might be like. Remember the things people need to stay alive: food, air, and water. How will the colonists get all these things? What kind of research will they do? What will they do for fun?

To find out more about research on Mars and the Mars mission, visit this website: www.spaceflight.nasa.gov/mars.

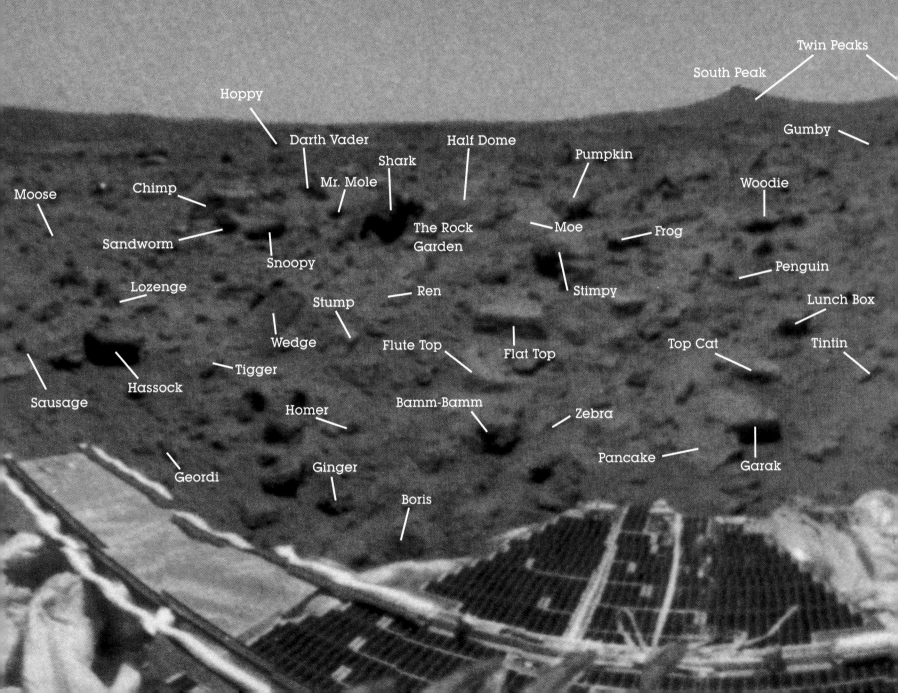

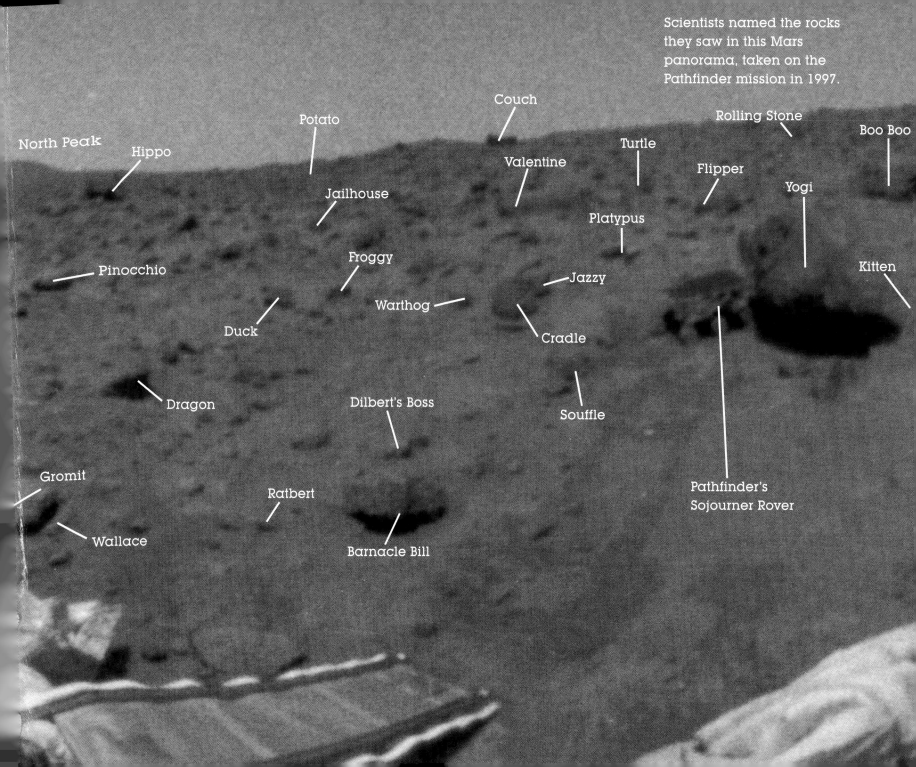

Scientists named the rocks they saw in this Mars panorama, taken on the Pathfinder mission in 1997.